STATS
TO
BLOW
YOUR
MIND

STATS TO BLOW YOUR MIND

13-Digit ISBN: 978-1-64643-224-0
10-Digit ISBN: 1-64643-224-X

This book may be ordered by mail from the publisher. Please include $5.99 for postage and handling. Please support your local bookseller first!

Books published by Cider Mill Press Book Publishers are available at special discounts for bulk purchases in the United States by corporations, institutions, and other organizations. For more information, please contact the publisher.

Cider Mill Press Book Publishers
"Where good books are ready for press"
PO Box 454
12 Spring Street
Kennebunkport, Maine 04046

Visit us online!
cidermillpress.com

Typography: Dazzle Unicase, Halcom

All vectors used under official license from Shutterstock.com.

Printed in Singapore

1 2 3 4 5 6 7 8 9 0
First Edition

STATS TO BLOW YOUR MIND

75 FACTS
THAT PROVE TRUTH IS STRANGER THAN FICTION

BY TIM RAYBORN

CIDER MILL PRESS

BOOK PUBLISHERS
KENNEBUNKPORT, MAINE

CONTENTS

INTRODUCTION

"FIGURES OFTEN BEGUILE ME, PARTICULARLY WHEN I HAVE THE ARRANGING OF THEM MYSELF; IN WHICH CASE THE REMARK ATTRIBUTED TO DISRAELI WOULD OFTEN APPLY WITH JUSTICE AND FORCE: 'THERE ARE THREE KINDS OF LIES: LIES, DAMNED LIES, AND STATISTICS.'"

MARK TWAIN,
"CHAPTERS FROM MY AUTOBIOGRAPHY"

It's doubtful that British statesman Benjamin Disraeli spoke these witty words (and Mark Twain certainly didn't come up with the phrase, even though it's often attributed to him), but there's definitely some truth to them. Statistics are used everywhere to sell things, prove points, win arguments, influence public opinion, and generally annoy people. Numbers and results can also be manipulated for dishonest and even nefarious purposes. But sometimes they can be entertaining, which is what this little book tries to do.

Herein, you'll find all sorts of unusual, amusing, and often-astonishing statistics about people and the world we live in, including everything from how many people are online to how many hamburgers a major fast-food chain sells every second. How many trees are there on planet Earth? The answer is: a lot more than you think. What are your chances of being killed by a vending machine? The odds are in your favor, but it does happen! How many times in your life could you walk around the world? Lace up your shoes and find out.

These numbers, of course, don't reflect the real world. If a study claims that 5 billion people use the internet every day, the researchers haven't actually polled 7 billion people and discovered that 5 billion of them are online. These numbers are extrapolations, using sample groups and assigning those numbers and results to a wider population. This is one reason some people feel that statistics aren't all that useful. Further, they can be made to say anything the researchers like. And there is some truth in this. It's easy enough to fudge numbers or only present certain results to make claims that the study backers want to hear. Companies do this all the time. Even when the studies are conducted by reputable organizations, they will often produce different results because there are so many other factors involved than simply asking people questions, such as who was polled, where the study was done, and when the research was conducted.

It's worth remembering that statistics are constantly changing and notoriously difficult to confirm or deny. These are general trends and averages, not absolutes. That's why you'll see many uses of words like "about," "approximately," and "as of 2021" throughout this book. Nothing is set in stone, and in five years (or even two), these numbers might be quite different. Except for things like the number of bones in your body; that's not going to change, unless you get bitten by a radioactive spider, or something. But otherwise, if past trends tell us anything, these numbers will probably change over time. The sources listed for each statistic are reliable, but even they get it wrong sometimes.

This book details some pretty amazing facts and figures, so enjoy dipping into the following pages and finding out that the world is a more amazing, wonderful, and weird place than you think. So are the people that live in it, but there's a 91% chance that you already knew that.

THE ONLINE WORLD

THERE ARE MORE THAN A HALF BILLION INSTAGRAM ACCOUNTS ACTIVE EVERY DAY

Being an influencer is more and more difficult; you have to shout to get heard, or you won't influence anything. With numbers like these, it's no surprise. Instagram has become a cultural phenomenon, at least for the moment, but many are already moving on to other platforms, such as TikTok. How long will this large number of Instagram users last? It depends on the site's ability to adapt and keep things fresh and interesting for people, always a challenge for a vast audience that's hungry for something new and gets bored easily.

= 1 million Instagram accounts

AS OF 2021, THERE ARE ABOUT 5 BILLION ACTIVE INTERNET USERS WORLDWIDE

That's over 60% of the world's population! It's stunning to think that something that, even only twenty years ago, was still a bit of a niche technology has grown to such astonishing numbers of users. The internet is arguably the most important piece of technology developed in the last century. Some even equate its significance to that of the printing press. Few foresaw in the 1990s how much it would come to dominate almost every aspect of world culture, but here we are.

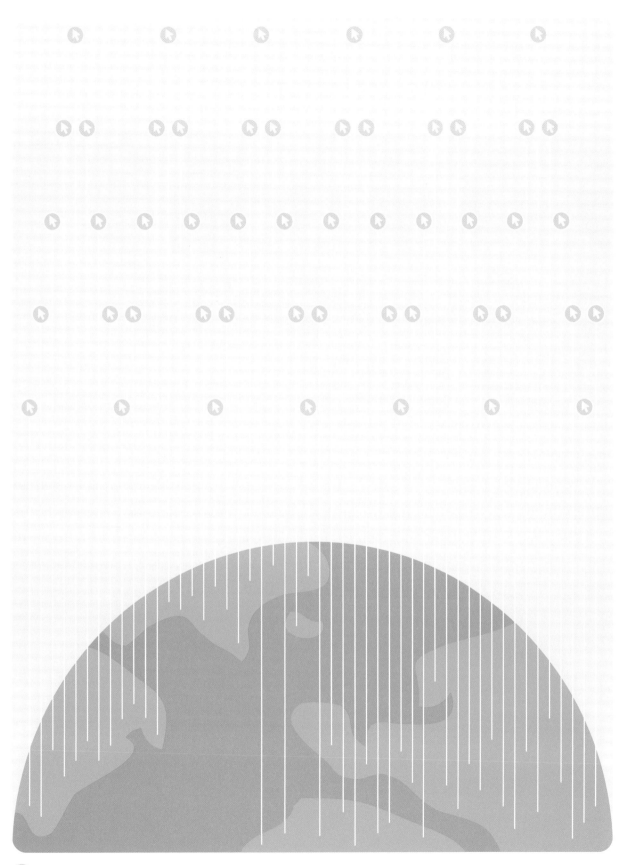

= 100 million active internet users

BLOGGERS PUBLISH ABOUT 7.5 MILLION BLOG POSTS PER DAY

Blogging is still one of the main ways that internet personalities—and even the non-famous—put down their ideas and stay in touch with followers. No matter what you're into, there's a blog (or ten) about it somewhere. So don't worry if you can't keep up with all the ones you actually follow. Blogs are just another part of that daily onslaught of information we have to wade through. Support the ones that are most meaningful to you, and ignore the 7,499,993 others.

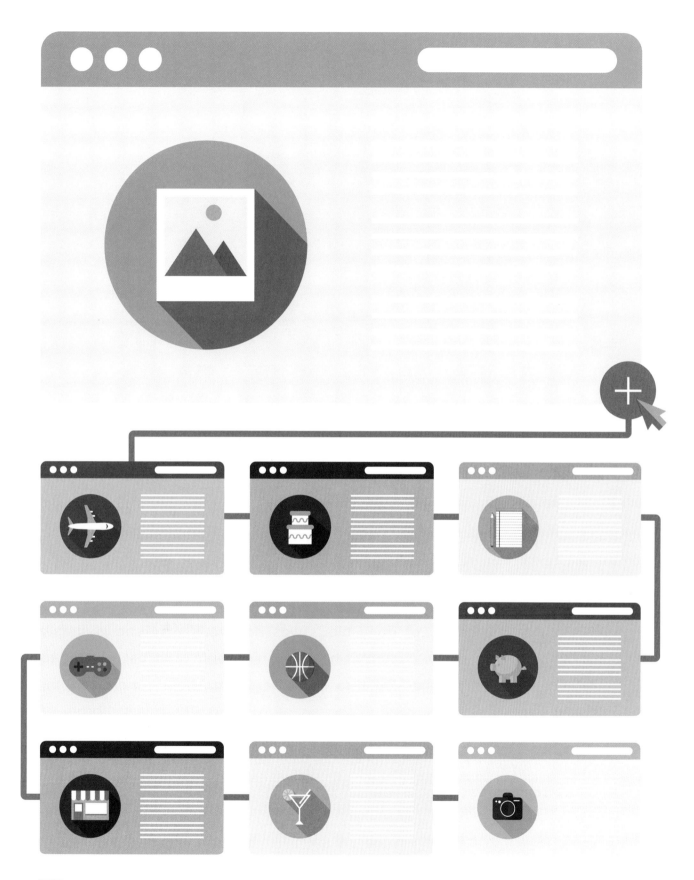

= 750,000 blog posts

AS OF 2021, MORE THAN 500 HOURS OF VIDEO ARE UPLOADED TO YOUTUBE EVERY *MINUTE*

If you thought the number of daily blogs was ridiculous, YouTube is simply staggering for the amount of content it holds. Many people report struggling to keep up with even the few channels they do subscribe to. There is more content on YouTube than could be viewed in thousands of years. It's a vast library of video unlike anything ever known. There's something for everyone there... and a lot that probably isn't for anyone! It can also take up a good chunk of your day. Use your watch time wisely.

 = 1 hour of video

ENGLISH DOMINATES THE WEB, WITH MORE THAN 25% OF ALL WEBSITES IN ENGLISH

25.3% of all websites are in English, followed by almost 20% in Chinese. For better or worse, English has become the preferred international language, and websites more and more reflect this. Even sites in another language frequently now have an English translation as an option. While Mandarin Chinese is still technically the language spoken by the most people worldwide as a first language, English dominates as a second language and for business communication. The internet will almost certainly continue to be "Anglicized," at least as far as language goes.

AS OF 2021, GOOGLE HOLDS ABOUT 92.7% OF THE SEARCH ENGINE MARKET SHARE

All the other search engines combined are so far behind in second place, they're hardly noticeable. And yet there are some great alternatives to Google. Many people are unhappy with Google's privacy issues, its monopoly-like dominance of the market, and its questionable activities in other areas. For a company that once had the motto "don't be evil" (but, significantly, has since dropped that), many would say they've strayed far from their ideals. But droves of people still keep going back to this search engine (with something like 5.6 billion searches per *day*!), simply because it is accurate, reliable, and easy to use. The fact that "google" is now a verb in most dictionaries shows how deeply imbedded it's become in our lives and thinking.

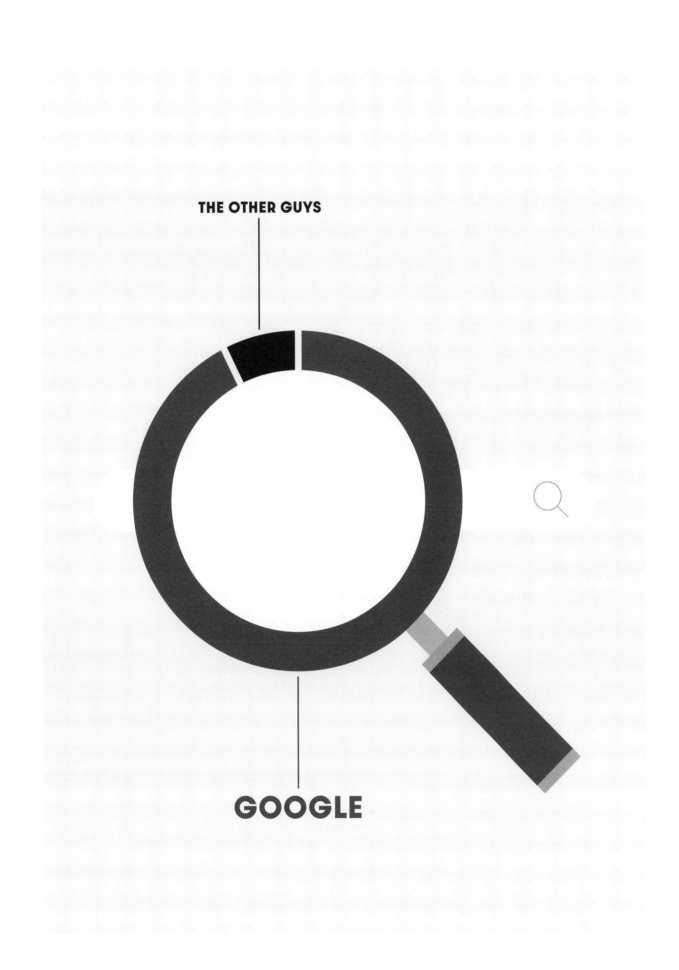

THE OTHER GUYS

GOOGLE

WORLDWIDE, ABOUT 4.28 BILLION PEOPLE USE THEIR MOBILE DEVICES TO GO ONLINE

That's more than 87% of all internet users! Once upon a time, PCs and laptops were our preferred way to get online and "surf the web." With the advent of phones that really are barely phones anymore, that's all gone out the (Microsoft) window. People use their phones to connect anywhere at any time, so it's not surprising that this is the preferred method to interact with the world. What may be even more surprising is that there are over 4 billion people around the world who have mobile devices of some kind, which shows just how pervasive they are.

👤 = 10 million people

AS OF 2021, THERE ARE WELL OVER 1.86 BILLION WEBSITES

At one point, there was a single website. Now look at how many there are. The number continues to grow, with up to 547,000 being added every day! No matter what you're into, there are websites for it, which is both a good thing and a bad thing. Even if our method of accessing them has changed (see page 24), websites still need to exist. It will probably be a long time before they go away, even though they've radically transformed from the simple (and quaintly humorous) sites of the 1990s.

= 10 million websites

AMERICANS RECEIVE OVER 4 BILLION ROBOCALLS PER MONTH

We've all been annoyed by these intrusions, which are just phone spam, but various "do not call" options have either proven ineffective or have expired, allowing this scourge to invade our lives even more than before. One wonders what the creators of these things think they're doing. No one wants them, no one cares, and often they're little more than scams. Hint: you car's warranty cannot be extended by a random phone call.

SUN	MON	TUE	WED	THU	FRI	SAT

! = 4 million robocalls

ACCORDING TO RECENT RESEARCH, BETWEEN 50% AND 85% OF ALL EMAILS ARE SPAM

That's a pretty broad range, but different studies produce different results, depending on country and time of year. But it's a lot, and your filters are still working overtime! This junk is never going away, sorry to say. Like robocalls, there will always be emails from a prince in Nigeria who wants to give you his fortune, a credit card company that needs you to "clink this link" to update your account information, and offers to teach someone's child music (or language, or whatever) in exchange for a certain amount paid in advance. They're all scams, and yet incredibly, after a quarter century of internet use, a lot of people still fall for them. Don't be one of those people.

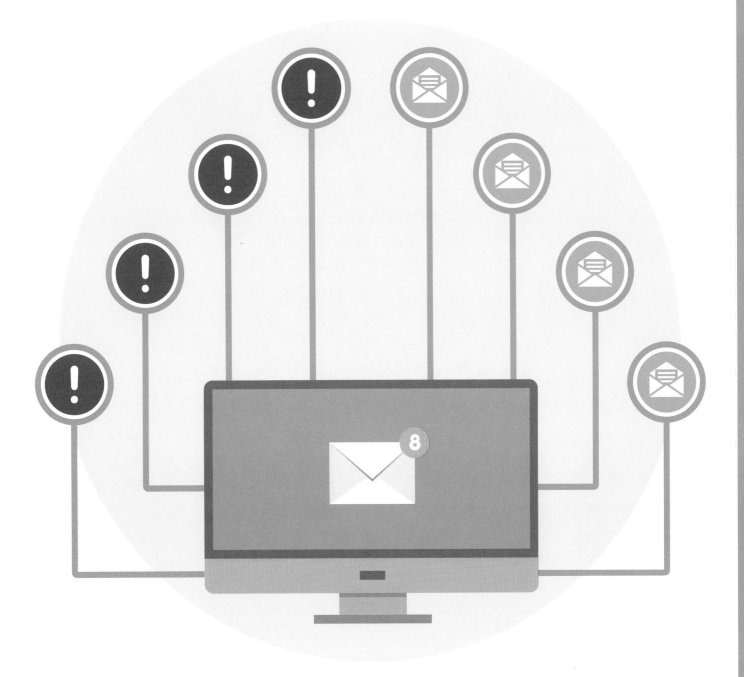

NETFLIX DOMINATES THE STREAMING INDUSTRY, WITH NEARLY 208 MILLION SUBSCRIBERS WORLDWIDE AS OF MID-2021

Multiply that number out by an average of $10 a month per user, and you can see how they keep funding all these expensive new shows and movies. And then canceling the shows you love after one or two seasons. But that's a completely different topic. Netflix was able to successfully pivot away from just offering an alternative to Blockbuster stores (remember those?) and is now a studio in its own right, producing award-winning movies and series with top stars. And their subscription numbers show no signs of decreasing any time soon.

= 1 million subscribers

ABOUT 1 IN 10 PEOPLE WHO ARE ON A ZOOM CALL ARE WEARING BUSINESS-APPROPRIATE SHIRTS, WHILE ALSO WEARING SWEATPANTS

It's the so-called "Zoom mullet." Fess up, we've all done it at least a few times: three-day-old sweatpants and bunny slippers with a pressed white shirt. Said sweats and slippers are carefully hidden out of view of your device's camera, and no one is any the wiser. We're all going away from the meeting thinking that we're so clever, but in fact, at least a few others at said meeting were probably doing it, too. 1 in 10 seems a bit low, to be honest, and is probably only a measure of the people willing to admit to it.

NEARLY 97% OF AMERICANS OWN A CELL PHONE OF SOME KIND; 85% OWN A SMARTPHONE

That's about 318 million people, an absolutely astonishing number; just about everyone in the U.S. is plugged into the Matrix now! Cell phones have become as common in American daily life as the internet, automobiles, and deep-fried Twinkies. These devices have countless useful applications beyond phoning, of course, and many now wonder how they ever did without them. Yet somehow, earlier generations didn't need to be attached to their phones for hours a day, so the mystery remains. Watching large crowds of people walking down the street, their faces glued to their phones and oblivious to their surroundings makes one wonder if we got ideas about the zombie apocalypse wrong.

NO CELL PHONE

BASIC PHONE

SMART-PHONE

NEARLY ONE-THIRD OF AMERICANS— 31%—SAY THAT THEY ARE ONLINE "ALMOST CONSTANTLY"

About half of all people surveyed say they go online several times a day. Perhaps not surprisingly, those under the age of 50 are the ones most often online. On the opposite side, only about 7% of people never use the internet at all. But these folks are a vanishing breed, it seems, and in another generation, it's likely that everyone will use the internet at least some of the time.

AS OF 2020, THERE ARE ABOUT 26 "SMART OBJECTS" FOR EVERY PERSON ON EARTH

This is about more than computers and phones. These days, watches, cars, clothing, refrigerators, and so much more are plugged into the web, making it hard to discern whether someone is ever truly offline.

EAT, DRINK, LIVE

A SINGLE ACRE OF LAND CAN GROW UP TO 50,000 POUNDS OF STRAWBERRIES

You could make a lot of jam with that! Or smoothies. So many of us are now so divorced from the land that feeds us that we're not even aware of how it's used and what it's capable of. We may enjoy growing herbs in clay pots on our windowsill, or have a small garden in the backyard, but the sheer magnitude of what farming is capable of is staggering. Only about 2% of the U.S. population works in farming, and the rest of us rely on their hard work.

= 100 pounds of strawberries

ON ANY GIVEN DAY, 1 IN 8 AMERICANS (OVER 41 MILLION PEOPLE) EATS PIZZA

A New York slice, Chicago Deep Dish, Sicilian . . . let the debates over which one is best begin! Is pizza Italian, or is it an American invention? It's both, of course, and it's become one of the most quintessentially "American" foods we have, thanks to certain nineteenth-century Italian immigrants who brought pizza traditions with them from Naples and came up with new ideas to improve this dish in U.S. cities. By the first decade of the twentieth century, pizza was being sold in eateries and stores in Italian neighborhoods nationwide. Its popularity grew over the next few decades, and after World War II, it was well on its way to becoming the phenomenon that it still is today.

THE FRENCH EAT 10 BILLION BAGUETTES PER YEAR

At about 29 million households in France, it works out to 347 baguettes per year per household, or slightly less than one a day. Still, it's an astonishing number, and also works out to people consuming about 320 long loaves of bread every second! The baguette is still so essential to French culture that in the pandemic, French bakeries remained open as essential businesses. Indeed, while people in other nations may have stockpiled toilet paper and hand sanitizer, the French were more fearful of a bread shortage.

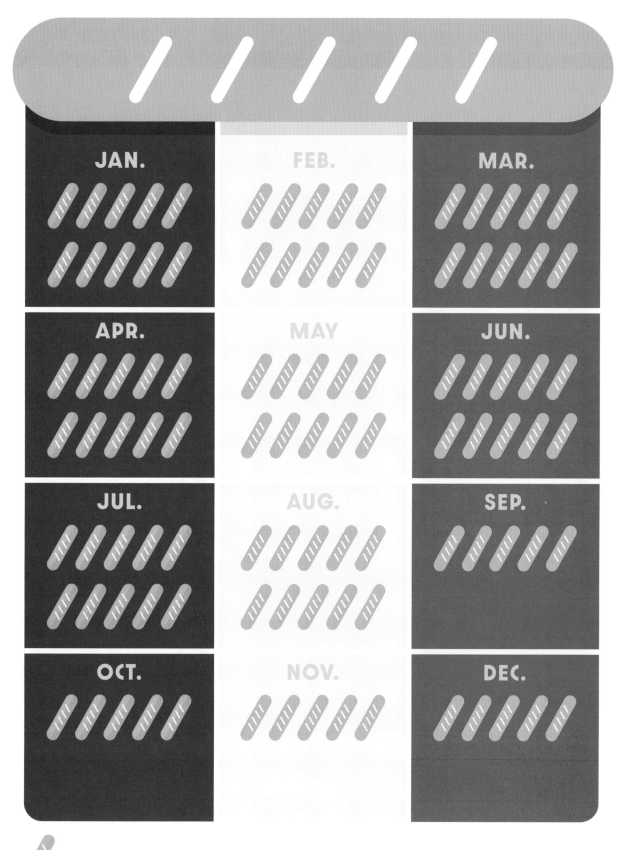

JAN.

FEB.

MAR.

APR.

MAY

JUN.

JUL.

AUG.

SEP.

OCT.

NOV.

DEC.

 = 100 million baguettes

CANADIANS CONSUME 1.7 MILLION BOXES OF KRAFT MACARONI EACH WEEK

About 7 million boxes are sold each week around the world, which means that Canada alone consumes almost 25% of all Kraft Mac & Cheese sold in that time! Canadians also eat 55% more of it than is consumed in the United States in the same period. Forget poutine or maple syrup, this easy-to-make food is a good candidate for being the national dish of Canada.

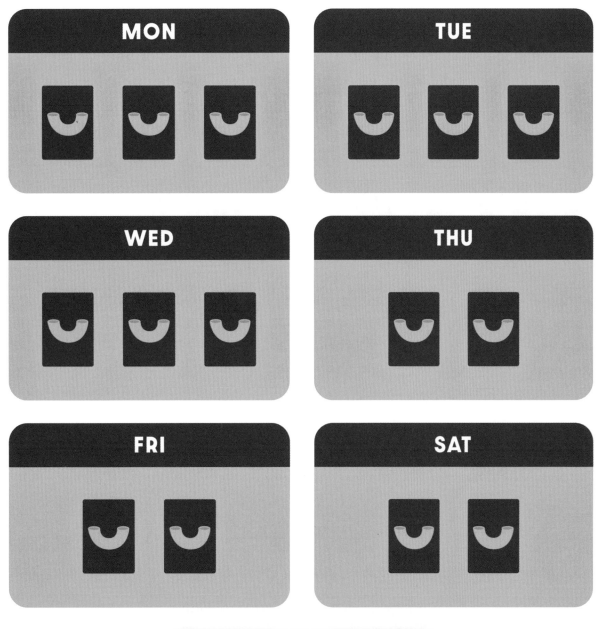

 = 100,000 boxes

THE AVERAGE RUSSIAN EATS ABOUT 7 POUNDS OF ICE CREAM PER YEAR

Now, Russia is a pretty cold country for many months at a time, but, surprisingly, ice cream consumption is just as popular in winter as in summer, if not more so. A famous legend tells of Winston Churchill going to Moscow in the winter of 1944 and seeing a group of Russians outside, in the freezing cold, eating ice cream. "These people will never be defeated," he is reported to have said. As of 2021, Russia is expected to produce about 510 tons of ice cream for its eager population, and the industry shows no signs of slowing down.

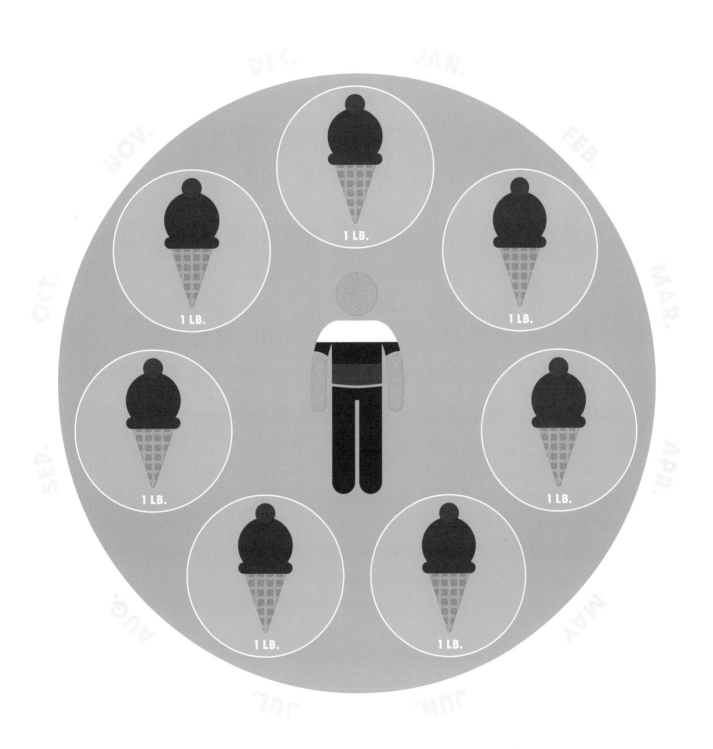

40% OF ALL FOOD GROWN AND PRODUCED IN THE U.S. IS NEVER EATEN

Food waste is a serious problem in the United States, and much more needs to be done to get that food to the people who need it. Distribution systems are either outdated or otherwise flawed, and grocery stores often have the policy of removing perfectly good food, such as produce, after a short period of time to make way for newer and fresher-looking food that will presumably be more appealing to consumers. Fortunately, some things are beginning to change, with more and more stores donating leftover food to charities and services that can use it. There are even produce delivery services that offer "flawed" fruit and veggies at a discounted price, produce that might look less than beautiful, but is still perfectly good to eat. These are small steps but essential ones to ensure that more food is eaten and not simply thrown away.

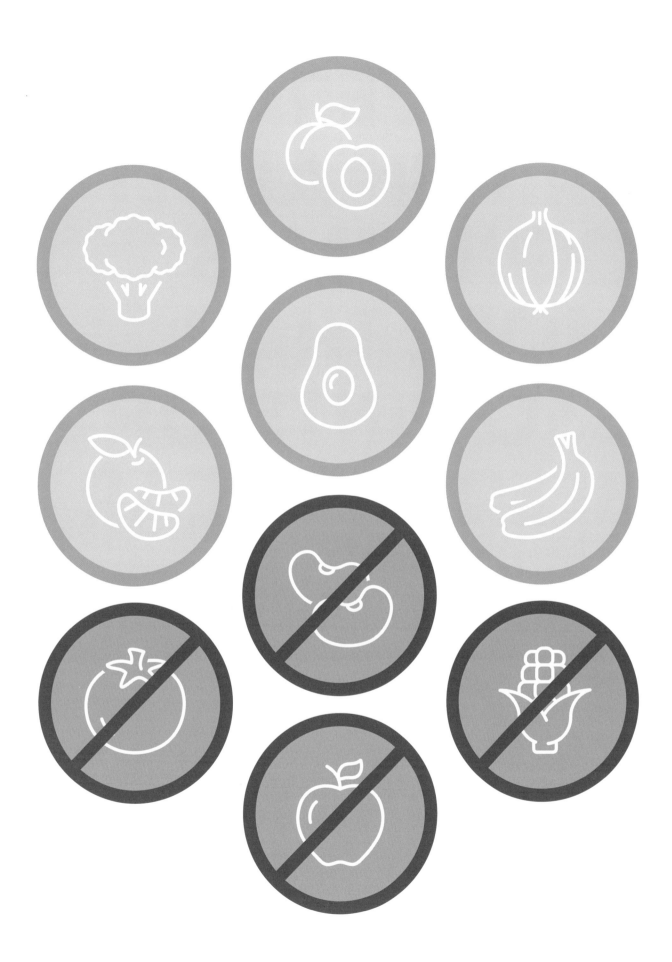

THE AVERAGE BRITISH ADULT CONSUMES MORE THAN 430 PINTS OF BEER PER YEAR

Statistical results range from the high 420s to about 440, but the number itself is rather shocking. It's also the equivalent of about 108 bottles of wine per year. This takes in both alcohol consumed at home and alcohol consumed elsewhere (such as at bars and pubs). Interestingly, those age 65 and over were more likely to drink more heavily than those in the 16 to 24 age range.

= 10 pints of beer

AMERICANS WHO ARE 21 AND OLDER (ABOUT 75% OF THE POPULATION) CONSUMED OVER 26 GALLONS OF BEER AND CIDER PER PERSON IN 2020

Americans love their beer! Despite the pandemic (or maybe because of it?), beer consumption was robust and healthy in 2020. Beer production is big business in the U.S., with the industry shipping the equivalent of over 2.8 billion cases of 24, 12-ounce containers in 2020 alone, along with 45.5 million cases of cider and similar drinks. Much of this was produced in the U.S. itself. Beer is booming and shows no signs of slowing down.

MORE THAN HALF THE WORLD'S POPULATION IN DEVELOPED COUNTRIES OWN A PET

Surveys show that some 57% of all people have pets of some kind. About one-third (33%) of these people own dogs, and nearly one-quarter (23%) own cats. Less popular, but still significant, pets include fish (12%) and birds (6%). Argentina has the highest number of pet owners (an amazing 82%), followed by Mexico (81%). The U.S. has the fifth-highest number, at 70%. Our furry (and other) friends are clearly very important to us, and the number of pet owners is growing all the time.

THE FRENCH SPEND ABOUT 40% LESS ON CLOTHING THAN THEIR NEIGHBORS IN ITALY

You would think with France being known for its high culture and fashion that its people would be shelling out the big euros for the latest in chic fashion trends and accessories. But it's not so. According to a 2016 study, the average French person in that year spent about 3.7% of their total expenditures on clothing, while in Italy the amount was 6.2%, In fact, France came in behind Portugal (6.3%), Germany (4.5%), and even Britain (5.6%). Not only was the number lower than these nations, it was also lower than the amounts spent in Bulgaria, Romania, and Hungary. It seems the French are more into bargain-hunting when they can, and can still work their fashion magic without breaking the bank!

THERE'S ABOUT A 50% CHANCE THAT YOUR LOST REMOTE CONTROL IS ACTUALLY STUCK BETWEEN YOUR SOFA CUSHIONS

This should be obvious, right? Usually the remote just slips down there, along with coins, half-eaten crackers, crumbs, and other things you probably don't want to think about. What this statistic really tells us is that we need to clean our sofas more often. Strangely, there's also about a 4% chance that said remote ended up in the refrigerator or freezer. And you have no one but yourself to blame for that.

DISNEY THEME PARKS HAVE MORE THAN HALF OF THE TOTAL THEME PARK ATTENDANCE IN THE U.S., AND ABOUT 55% OF TOTAL ATTENDANCE AT THE TOP 20 AMUSEMENT PARKS IN NORTH AMERICA

Disneyland and Disney World are big draws, so big that they dominate the theme park market. Attendance in 2018 was recorded at 157,311,000 for Disney attractions. The next largest attendance was for Merlin Entertainments Group, which owns Legoland, Madame Tussauds, and other attractions around the world. But this number was "only" 67,000,000, not even close to Disney. It looks like the House of the Mouse is not getting off the champion's podium any time soon.

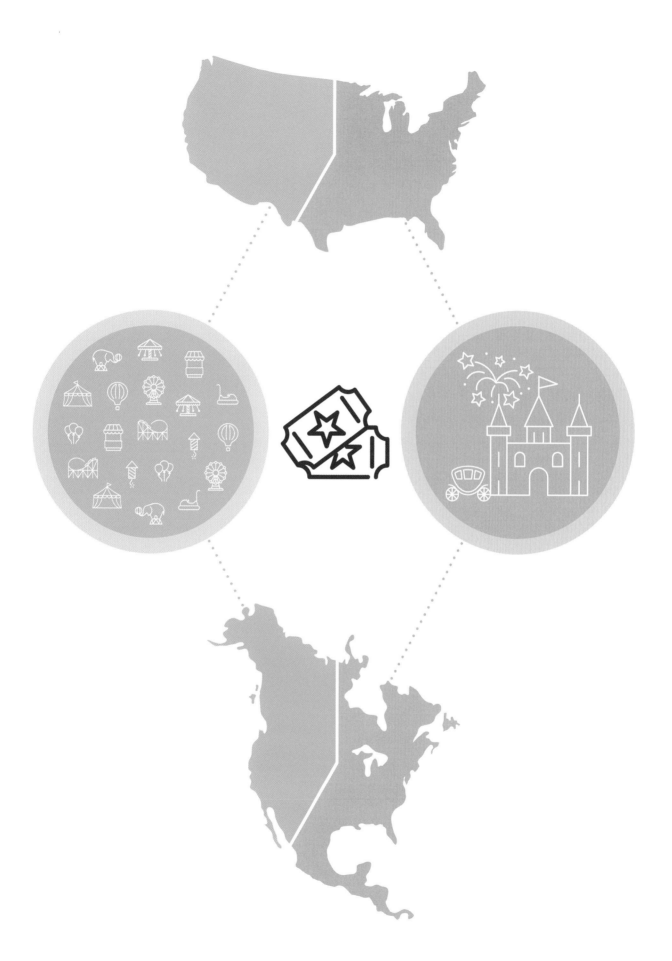

ONE-THIRD OF AMERICAN ADULTS STILL SLEEP WITH A "COMFORT OBJECT" FROM THEIR CHILDHOOD

34% of people, to be precise. We all probably had our favorite teddy bear, other stuffed toy, or blanket as children, the one that kept the monsters away and allowed us to get a good night's sleep. It turns out that, as adults, quite a few of us still need that comfort and security. We tend to be chronically sleep-deprived; only 27% of people say they get a good night's sleep on a regular basis. This is due to many causes, including our nonstop culture, the need to be plugged-in all the time (studies show that lighted screens disrupt sleep), the stresses of daily life, and a host of other causes. So if cuddling a cherished childhood toy is a way to calm someone down and allow them to relax and rest better, more power to them. Yay for teddies!

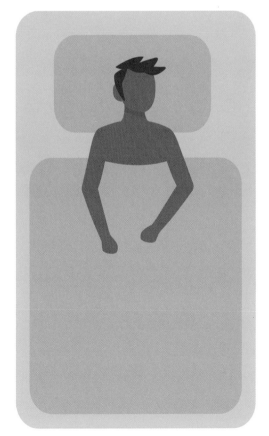

40% OF THE WORLD'S CALORIES COME FROM JUST THREE CROPS: RICE, WHEAT, AND MAIZE

These plants are remarkably versatile, which has allowed them to be grown all over the world, but this is a wasteful and even potentially dangerous use of resources. There are at least 30,000 edible plants in the world, of which humans have cultivated between 6,000 and 7,000. Of these, only about 170 are in regular use, and when it comes to staples, we're back to rice, corn, and wheat. But what happens if a disease or blight wipes out one of these three? Our global food supply would be sent into chaos. It's time to look at the many other wonderful plants out there and start using more of them.

AS OF 2020, THERE ARE MORE THAN 39,000 MCDONALD'S SPREAD OUT ACROSS THE WORLD

McDonald's still dominates the fast food market and has stores on every continent except Antarctica; give them time and they'll probably figure out how to set up one there, too, maybe offering fish sandwiches to penguins? The company has increased its number of locations every year for at least 16 years, and shows no signs of stopping. Despite this market dominance, there are actually about 90 nations worldwide that don't have a McDonald's, many of them in Africa and Asia.

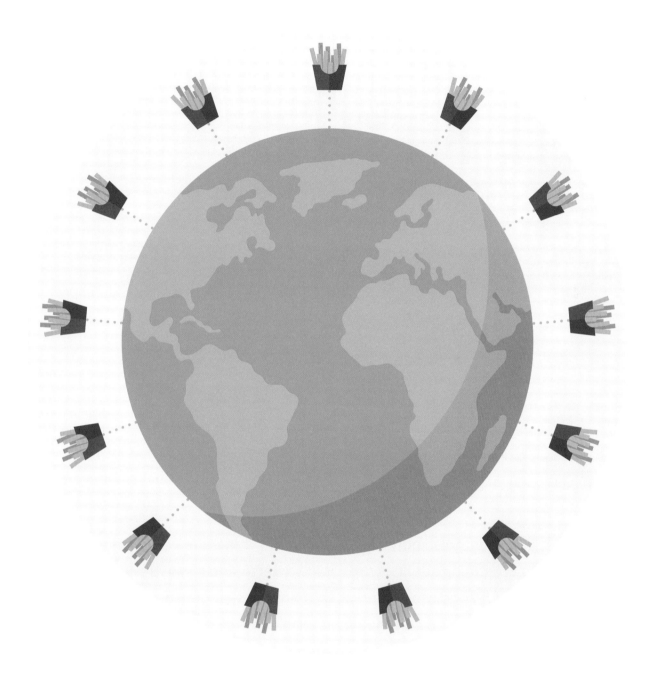

 = 3,000 McDonald's

PEOPLE IN THE U.S. SPEND ABOUT ONE-FIFTH OF THEIR TIME (ABOUT FOUR AND A HALF HOURS) EVERY DAY WATCHING TELEVISION

This seems like an astonishingly high number, but it includes on-demand TV and video viewing. Streaming services are rapidly overtaking terrestrial and cable TV services in popularity and usage, and as that happens, people are happier to binge things on their own time and on their own terms, rather than having to watch at a preselected time or record something for later. Cable companies have responded by modifying their own services to give viewers more flexibility, but it remains to be seen if they can keep up with Netflix, Disney+, and the other streaming services now taking over the world of entertainment.

MCDONALD'S SELLS MORE THAN 75 BURGERS EVERY SECOND

This number seems incredible, until you look at the number of restaurants on page 72 and realize that this chain is so widely spread out. That's almost 6.5 million burgers a day around the world. Actually, this figure is a few years old, so it's entirely possible that the number is even higher now.

CONSUMERS IN TURKEY DRINK MORE TEA PER PERSON THAN ANY OTHER NATION

You might think that China would hold this honor, or even England (given its love affair with the drink), but no, on a per-person level, the Turks consume about 6.96 pounds (3.16 kilograms) each per year. Ireland is the nearest challenger, drinking up 4.83 pounds (2.19 kilograms) in a year. The Brits come in far below the Turks, with 4.28 pounds (1.94 kilograms) per person each year. China produces way more tea, of course, but when spread out over its massive population, the number comes to about 1.25 pounds (0.57 kilograms) per person every year.

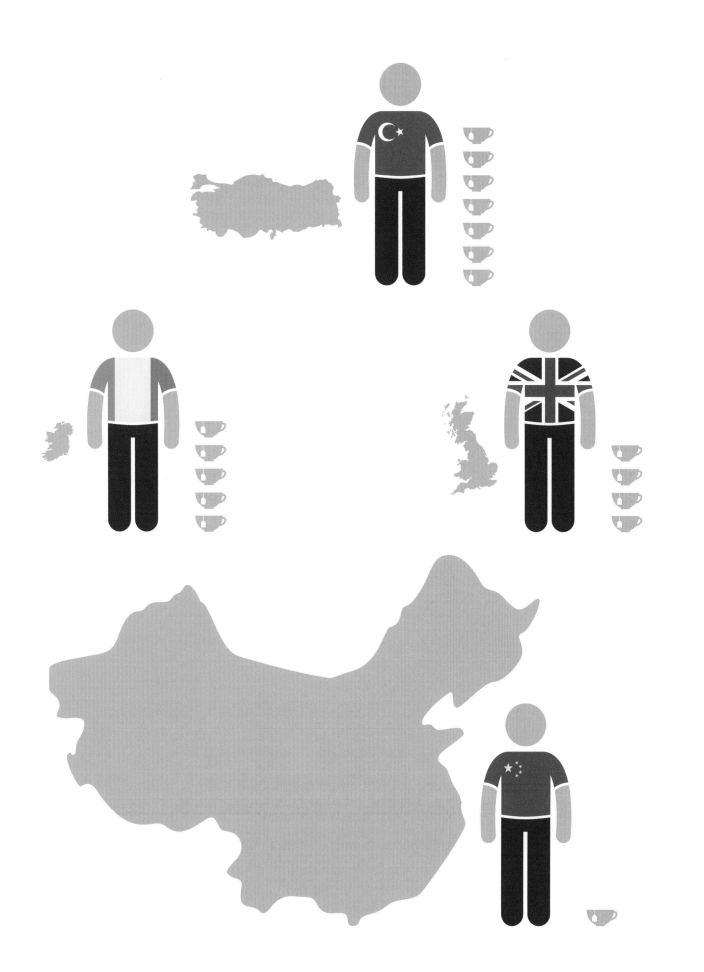

ABOUT 70 MILLION PEOPLE IN AMERICA CHECK THEIR HOROSCOPES ON A REGULAR BASIS

In addition, it's estimated that about one-third of Americans believe in astrology, or at least that there's something to it. This makes the scientific community roll its collective eyes, but it's very hard to remove entrenched beliefs of any kind. For most folks, it's just a fun little peek into how their day or week might go, and they don't really get into all of the confusing details and minutiae of it. If you like to have a quick look at your daily horoscope, don't feel bad; you're in good company.

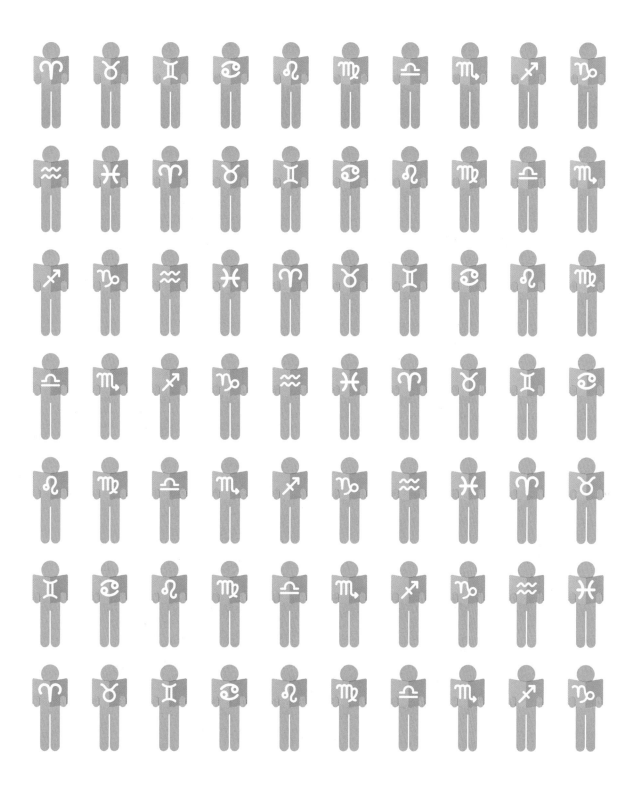

= 1 million people

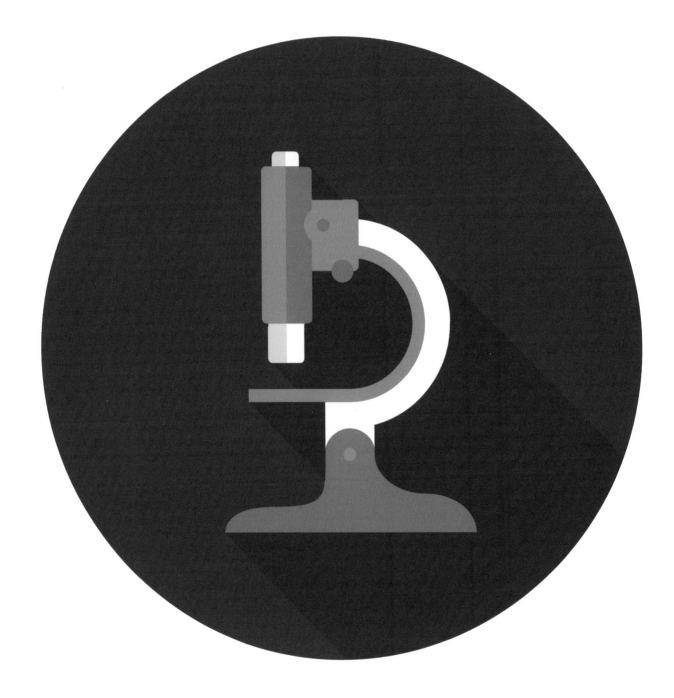

SCIENCE

THE HUMAN EYE BLINKS ABOUT 4,200,000 TIMES A YEAR

That's about 12,000 times per day! And some people blink much more than that, with estimates of more than 20,000 times a day. We blink constantly when we're awake. Blinking does all sorts of important things: it clears dirt and debris from your eyes, keeps them moisturized, and delivers nutrients. Our eyelids protect us from excessive light and damage, too. Studies show that our brain might consider blinking to be something like a "micro nap," giving the brain a chance to reset, which may even alter our perception of time. And you just thought it was something to clear your eyes!

= 100,000 blinks

WE BREATHE IN AND OUT BETWEEN 22,000 AND 25,000 TIMES A DAY

Obviously, those who engage in rigorous exercise might breathe more, while those with a more relaxed view on life might breathe less, but there's no doubt that we breathe a whole lot! Obviously, breathing is how we stay alive, and the process is kind of miraculous, as we take in air and send oxygen to our systems, while removing carbon dioxide from our bodies. But it needs to be done constantly, and it continues automatically in our sleep. Fun bonus fact: our lungs are different sizes; the left lung is a little smaller to accommodate the heart.

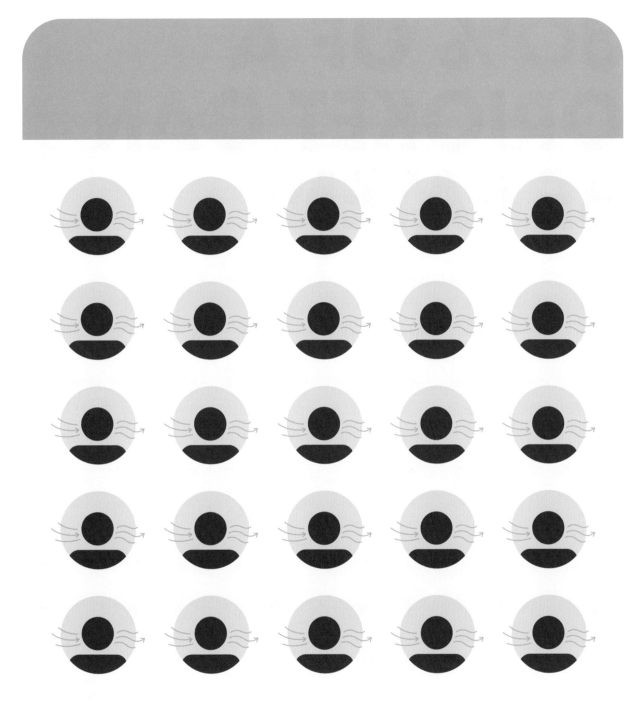

 = 1,000 breaths

80% OF A CRICKET CAN BE EATEN

When looking for high-protein alternative food sources, insects are seen by many as the best bet for the future. In contrast to poultry and pigs (55% of which can be eaten) and cows (40% of which can be consumed), insects seem to offer a better value for money and produce a lot less waste. They are also much quicker to grow and less fatty and unhealthy. The problem, of course, is consumer squeamishness, and convincing populations in many countries to give them a try. Still, insect snacks are already popular in many South American, African, and Asian countries, and might point to future trends in global eating.

LIGHTNING FLASHES EARTH ABOUT 44 TIMES EVERY SECOND

We tend to think of lighting as something intense and violent that happens during thunderstorms. While this is true, most people probably aren't aware of just how often lighting strikes. Somewhere over the Earth, lighting flashes 44 times a second, which works out to an incredible 1.4 billion times a year! One could say that Thor or Zeus is very busy! It's an uncomfortable reminder of just how active our atmosphere is, even on days that seem calm.

JUST 0.3% OF SOLAR ENERGY FROM THE SAHARA DESERT WOULD BE ENOUGH TO POWER THE WHOLE OF EUROPE

The sun gives off an incredible amount of energy, far more than we as a world civilization would ever need. The difficulty has been in capturing it efficiently, even the small amounts that would power everything. But this statistic shows how little of the sun's energy would be needed to take care of a huge energy need, both locally and around the world. With the increasing efficiency of solar batteries, it seems obvious that this rich, endless source of energy needs to be tapped into far more.

0.3%

NEW PARENTS LOSE ABOUT SIX MONTHS OF SLEEP IN THE FIRST TWO YEARS OF THEIR CHILD'S LIFE

New parents are well aware of the fact that they're basically not going to get a good night's sleep for a long, long time, if ever (the teenage years don't do much to improve this, it must be said). But it's actually worse than they feared. With being up multiple times over the course of the night, getting a long, uninterrupted sleep is not likely to be an option. Indeed, something like 60% of these poor parents get only about 3.75 hours of sleep per night. An additional 10% of people get only get about 2.5 hours of sleep each night. Clearly, this is bad for their health in the short term and long term, and it's essential to try to find other times to catch up with naps.

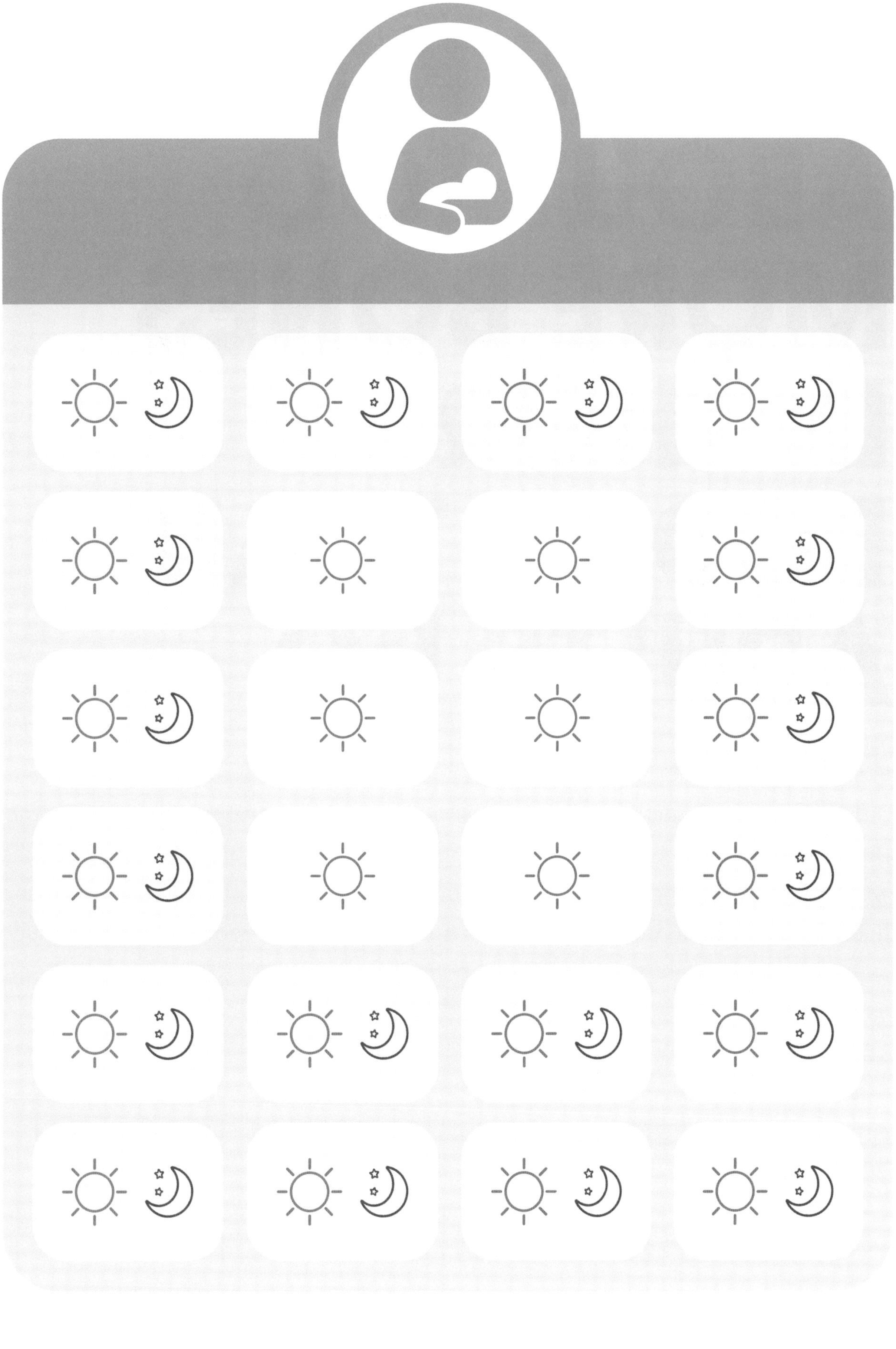

BABIES HAVE ABOUT 100 MORE BONES IN THEIR BODIES THAN ADULTS DO

Wait, how is this possible? The answer is that babies have many bones made almost entirely of cartilage; this allows them to curl up comfortably in the womb. But as infants begin to grow after birth, these bones will start to calcify and fuse with those around them. Thus, these nearly 300 bones eventually fuse down and harden into the 206 bones we have as adults. This is one reason calcium is such an important nutrient for babies and young children.

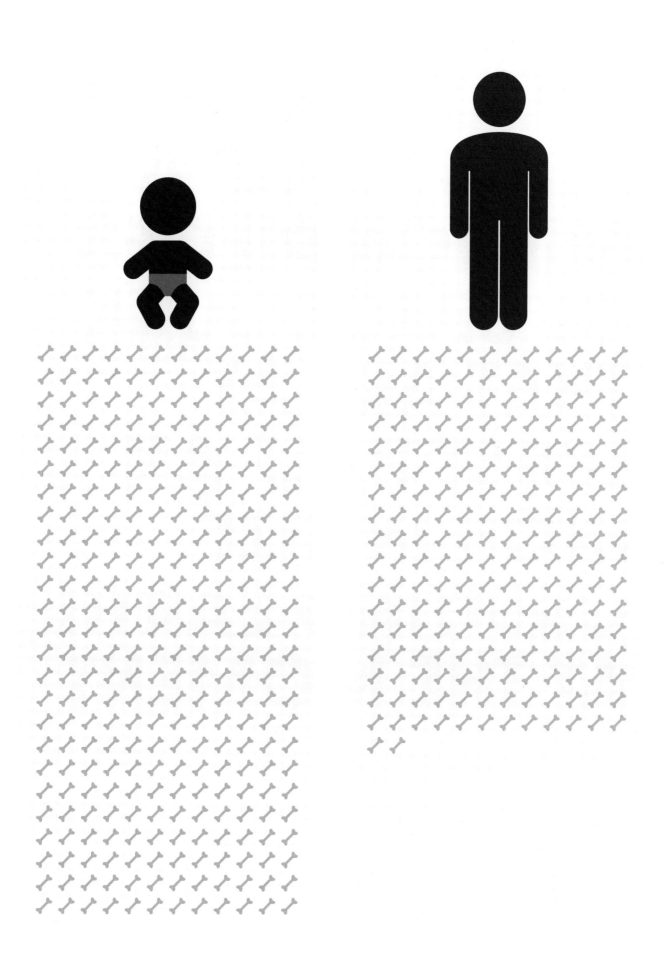

THERE ARE ABOUT HALF A MILLION PIECES OF SPACE JUNK THAT ARE AT LEAST HALF AN INCH IN SIZE ORBITING EARTH AT THE MOMENT

About 21,000 of these orbiting objects are larger than 4 inches in size. While most of these are not a danger to anyone on Earth (the odds of being harmed by a piece of space junk are about 1 in 100 billion!), they are traveling fast enough that they can cause considerable damage if they happen to slam into anything, such as a space shuttle, or the International Space Station. Fortunately, space is a big place, even close to Earth, so such collisions are rare and unlikely.

⊗ = 10,000 pieces

MORE THAN 38% OF RESPONDENTS AROUND THE WORLD BELIEVE THE (DEBUNKED) CONSPIRACY THEORY THAT VARIOUS GOVERNMENTS INVENTED THE COVID-19 PANDEMIC TO SOMEHOW CONTROL THEIR CITIZENS, OR CULL THE POPULATION

This should seem like an absurd idea (and it is), but unfortunately, such ideas are gaining more and more popularity thanks to the ever pervasive lies on the internet, where the only qualifications for posting anything are a connection and at least a partially functioning keyboard or phone. Disturbingly, younger people are more likely to believe conspiracy theories like this than older ones, which perhaps shows the level of distrust that younger generations have in systems that have indeed screwed up a lot of things and given young people good reasons not to trust them.

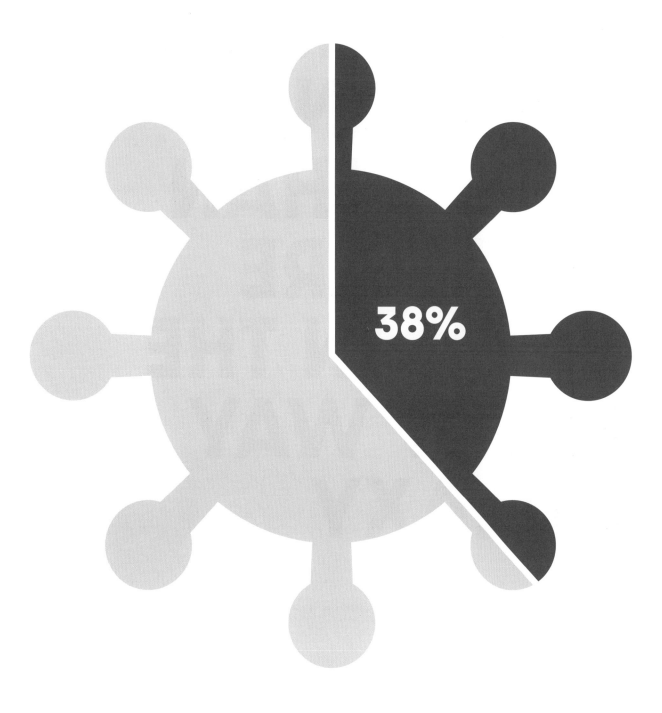

THERE ARE MORE TREES ON PLANET EARTH THAN THERE ARE STARS IN THE MILKY WAY GALAXY

There are about 3 *trillion* trees on our planet, but "only" 100 billion to 400 billion stars in the galaxy. That's an astonishing number of trees (about 422 for every person on Earth), but incredibly, it's not as many as there once was. In the good old days before humanity, there may have been as many as twice that number. It's good news for those worried about world deforestation, but it also shows that we have had a significant impact, and should work harder to reverse that decline, since trees are very much the planet's lungs.

🌳 = 10 billion trees

✦ = 10 billion stars

ONE-QUARTER OF YOUR BODY'S BONES ARE LOCATED IN YOUR FEET

Each foot has 26 bones, so that's 52 for both, out of a total of 206 bones in your body. So, basically 25% of all of your bones are in your feet. Feet have to be pretty strong and flexible to do this inconvenient thing called standing up, so we're fortunate that we have such a complex and amazing system to help us do it. Perhaps just as amazing, we have 27 bones in each hand, so a full half of the bones in our body are just in our hands and feet. But think about all the amazing things they can do!

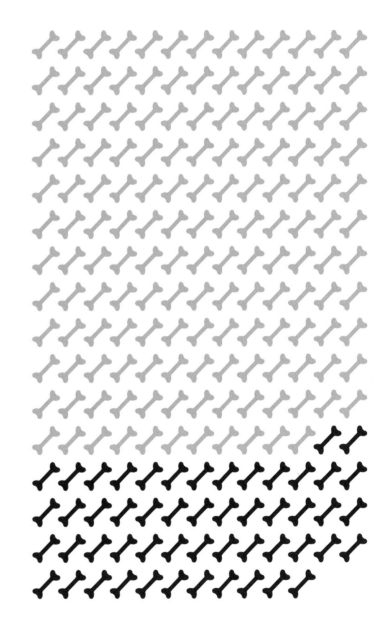

GLOBAL FACTS AND FIGURES

THE U.S. HAS THE WORLD'S LARGEST IMMIGRANT POPULATION, WITH 20% OF THE WORLD'S IMMIGRANTS

That number is based on a study from a few years ago and almost certainly is growing all the time. The U.S. has long been a destination for newcomers from all over the world, and while the subject of immigration can get contentious from time to time, with different political ideologies using it rather like a football, there is no reason to see this number decreasing significantly in the coming years. The reality is that issues like climate change, political instability, and water shortages may result in an increase in immigration numbers.

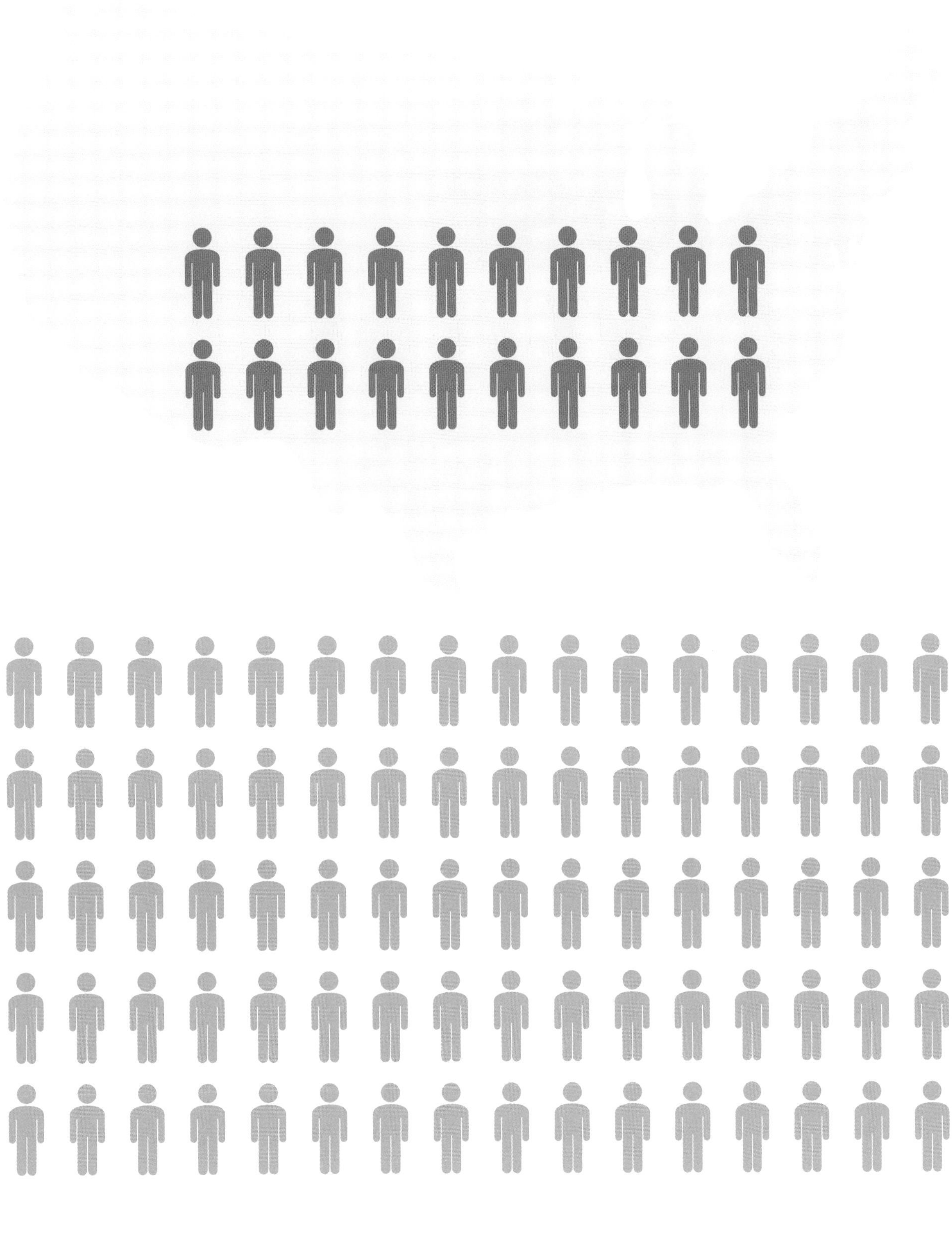

THE CANADIAN FLAG IS ALMOST 100 YEARS YOUNGER THAN THE NATION ITSELF

Canada became a self-governing nation within the British Empire on July 1, 1867, but it didn't adopt the instantly recognizable red-and-white maple leaf flag until the end of 1964. The current design was one of three finalists. The other two possible flags featured a three-leaf design bounded by dark blue, and the same design as the current flag, but with a Union Jack on the left panel and the French fleur-de-lis on the other panel, to symbolize the nation's British and French heritage.

1970

1965

1960

1955

1950

1945

1940

1935

1930

1925

1920

1915

1910

1905

1900

1895

1890

1885

1880

1875

1870

1865

1860

IT TAKES 68 DAYS TO SWIM THE ENTIRE LENGTH OF THE MISSISSIPPI RIVER

Yes, this was actually done by someone, a marathon swimmer named Martin Strel. He started from northern Minnesota on July 4, 2002, and completed his journey at the Gulf of Mexico in Louisiana on September 9, 2002. He swam every day for 68 days, averaging over 34 miles (55 kilometers) a day. Note that the typical mere mortal is not going to be able to do this feat in that same amount of time! Indeed, Strel received a Guinness World Record certificate for his efforts. A more realistic swim came from Chris Ring, who swam 6 days a week, and averaged about 15 miles (24 kilometers) a day. It took him six months. Either way, it's a heroic accomplishment!

FINISH

1 DAY ——

START

THE WORLD'S LARGEST NATIONAL PARK IS BIGGER THAN 166 COUNTRIES

Northeast Greenland National Park is a vast stretch of land covering about 375,300 square miles (972,000 square kilometers). That makes it larger in area than all except 29 of the world's nations. Despite its vast size, less than 50 people usually live there at any given time, mostly seasonal inhabitants such as police, park rangers, members of scientific expeditions, and military personnel.

UNITED
KINGDOM

FINLAND

NORTHEAST
GREENLAND
NATIONAL PARK

THAILAND

GREENLAND

ITALY

EGYPT

THE NATION OF BANGLADESH (ROUGHLY THE SIZE OF MICHIGAN) HAS A LARGER POPULATION THAN RUSSIA

This stunning stat is absolutely true: as of 2021, the population of relatively tiny Bangladesh in South Asia was approaching 165 million, while in the vastness of Russia (one of the largest nations on Earth in terms of square miles), the population was a bit under 146 million.

RUSSIA

BANGLADESH

👤 = 1 million people

2 OUT OF 5 AMERICANS CAN'T NAME ONE FREEDOM PROTECTED BY THE FIRST AMENDMENT

This is another of those concerning statistics that seems to come out of American research on a regular basis. About 40% of people surveyed couldn't name a single freedom offered by the First Amendment of the Constitution, while 9% of respondents thought that the right to bear arms was one of them. 36% of people could only name one freedom. Of those who correctly answered, freedom of speech was the most commonly cited freedom, followed by freedom of religion and the press, but the latter two were only named by 15% and 13% of respondents, respectively. These are worrying numbers. Understanding rights as citizens should be a basic given, but it's clear that, once again, the educational system seems to be failing many.

THE SUMMIT OF MOUNT EVEREST IS ROUGHLY THE SIZE OF TWO PING-PONG TABLES

At least according to some people who have been there. You know, just in case you feel like dragging a table up there and having a game or two. That does seem rather small for such a tall peak, but then again, mountains do have pretty pointy tops.

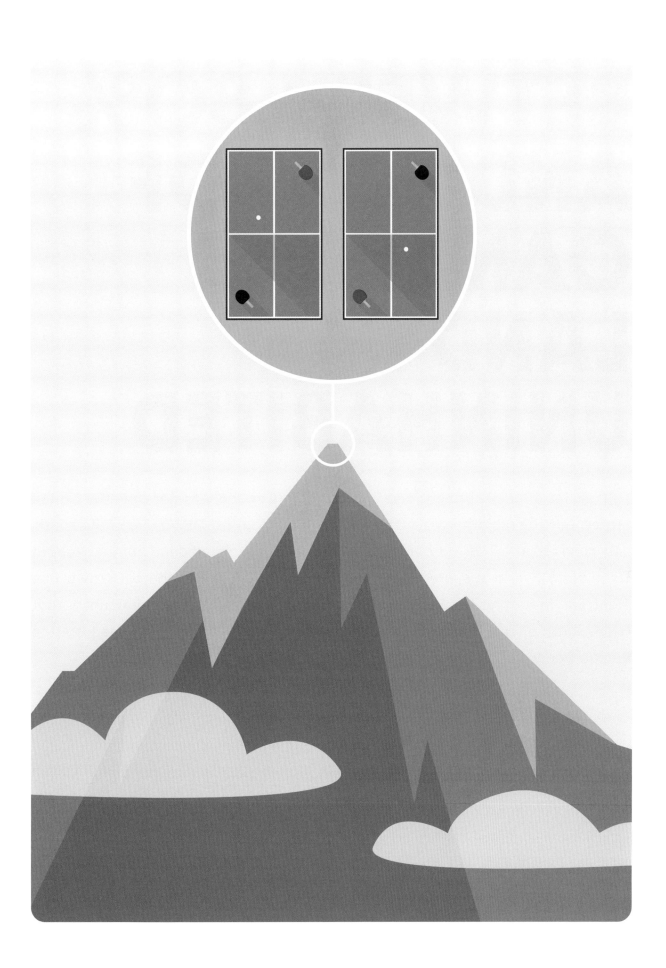

BETWEEN 80% AND 90% OF CANADA'S LAND IS UNINHABITED

That's a lot of uninhabited space! In total area, Canada is larger than the United States, but the majority of its 38 million people (almost 90%) live in the south, and close to the U.S. border. And of that 90%, a significant number (about 18 million) live in the area from Detroit to Toronto, and up to Ottawa and Montreal. In fact, about 50% of all Canadians live in areas of the country that reach south of the main straight U.S.-Canadian border that stretches from the Pacific Ocean to the Great Lakes.

80%

20%

IF YOU ARE MODERATELY ACTIVE, OVER THE COURSE OF YOUR LIFE YOU WALK THE EQUIVALENT OF 4.5 TIMES AROUND THE WORLD AT THE EQUATOR

You've probably heard that taking 10,000 steps a day is great for your health. An average person takes about 7,500 steps a day and lives about 80 years. Using these numbers, we can calculate that someone could walk about 111,850 miles (180,005 kilometers) during their lifetime, or more than four times around the world. Note: Americans as a whole tend to walk much less than people in other countries (it's all those cars!), often no more than 2,500 steps a day, so it would be fewer trips around the world for them.

IN THE U.S., EACH PERSON GENERATES UP TO 5 POUNDS OF GARBAGE EVERY DAY

The exact number varies a little, between 4.4 and 4.9 pounds (approximately 2 kilograms), but it adds up to a lot of waste. That's between 1,600 and 1,800 pounds (725 and 816 kilograms) of trash per person per year. And unfortunately, it's only getting worse. It seems that despite our efforts in recycling and trash reduction, the numbers keep growing as the population does. Only about half of all paper products are recycled, and packaging itself accounts for about one-quarter of all the waste generated. This stuff doesn't easily break down and can sit in landfills for decades, even centuries. The message is clear: we need to do better at reducing our waste.

THE POPULATION OF EARTH IS APPROXIMATELY 27 TIMES LARGER NOW THAN 1,000 YEARS AGO

There are different estimates about how many people were on Earth in the year 1000, but an average of around 290 million is probably not far off. Given that the current population is about 7,875,000,000, the math works out to 27 times more people. 1,000 years prior, in the year 1 CE, the population was between 270 and 300 million, so there was almost no growth over that thousand years, or maybe even a slight reduction, which is kind of mind-boggling to think of in itself.

YEAR 2021

7,875,000,000 PEOPLE

YEAR 1000 290,000,000 PEOPLE

ABOUT 385,000 BABIES ARE BORN AROUND THE WORLD EACH DAY

That's more than 140 million new arrivals every year. Conversely, about 150,000 people die every day, which adds up to more than 54 million per year. As you can see from these numbers, this means that the population is steadily growing, since only a little more than a third of the number of people die each day as are born. There are many different predictions about when and if the population will stabilize or even decline, with some saying it could happen as soon as the 2040s, or as late as about 2100.

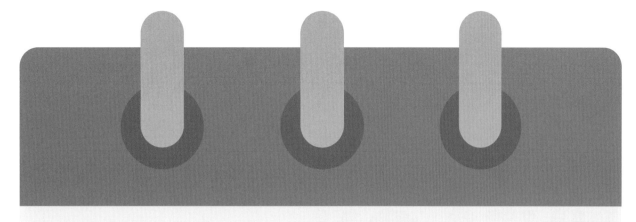

= 1,000 babies

ABOUT 95% OF EGYPT'S POPULATION LIVES ON APPROXIMATELY 4% OF THE LAND NEAR THE NILE RIVER

Egypt is the third most populated country in Africa, currently at about 100 million people. So, about 95 million people live very close to the Nile; this fertile area that's about half the size of Ireland has been the source of life and civilization for thousands of years. For comparison, the current population of Ireland is just under 5 million people, which should give you a better idea of the population density of Egypt's habitable land. Cairo clearly contains a good portion of these inhabitants, but other major cities and settlements dot the Nile from the delta all the way to Egypt's southern border.

95%
OF POPULATION

2,900,000 PEOPLE FLY IN AND OUT OF AMERICAN AIRPORTS EVERY DAY

This was not true during the height of the pandemic, of course, when many flights were grounded and people stayed home out of safety concerns. But Americans definitely love flying, and given the size of the country and the lack of other viable means of quick transportation, air travel it is. This number also includes passengers from other countries (visiting or just passing through), and it all shows how busy American airports are.

= 100,000 people

MORE PEOPLE LIVE IN THE NEW YORK CITY METROPOLITAN AREA THAN IN 46 OUT OF THE 50 STATES

This is another of those stats that makes you think, "That can't possibly be true!" But it is. About 19 million people reside in the greater NYC area as of 2021. Only California, Texas, Florida, and New York State itself have larger populations statewide. Pennsylvania, with about 12,800,000 people, comes in a distant fifth.

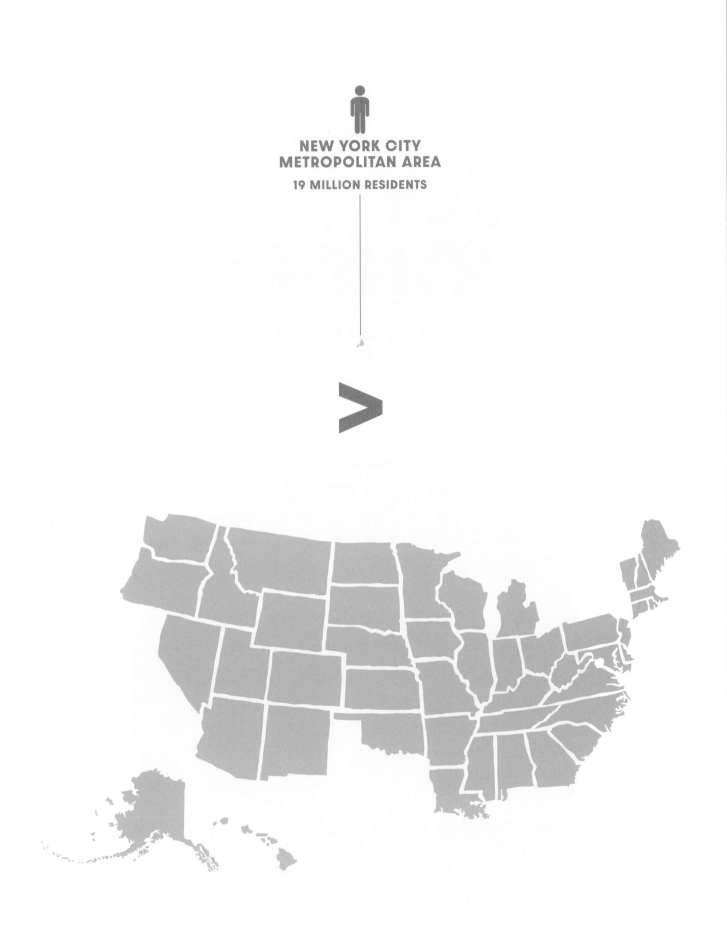

NEW YORK CITY METROPOLITAN AREA

19 MILLION RESIDENTS

>

11% OF U.S. SURVEY RESPONDENTS HAVE NEVER LEFT THEIR HOME STATE

This was from a single survey, but it is rather amazing that 1 in 10 Americans have never left their home state. The same poll also found that 40% of respondents had never traveled outside of the U.S., and about half have never owned a passport. This would tend to reinforce the idea of American insularity. But while there are doubtless many incurious individuals among those polled, a good number of those polled also indicated that they would like to travel, or travel more, but finances and work/family responsibilities often prevent it.

ONLY 31% OF PEOPLE IN AMERICA KNOW SOME OR ALL OF THEIR NEIGHBORS

As might be expected, the number is higher among rural residents (40%) and lower among urban residents (24%). Despite the numbers of people, it's actually easier to be anonymous in a city. Among throngs of people we can choose who we wish to interact with, and it often isn't the person next door or across the hallway. In rural areas, you often have to rely on someone down the road for a favor, for security, or whatever else you might need. The people in a small town will always know more of your business than those in a big city, whether you want them to or not.

ABOUT 22% OF AMERICANS SPEAK A LANGUAGE OTHER THAN ENGLISH AT HOME

This includes both citizens and residents. While the U.S. has no official language (despite what some might claim), English is obviously the lingua franca for a majority of the people living there. But at home, over one-fifth of the population is content to speak a different native language. These numbers are largest in the western states (from California to Texas), as well as Florida and New York, and the main second language is Spanish, followed by Chinese (Mandarin and Cantonese), Tagalog, and Vietnamese. Arabic, French, and Korean round out the top spoken second languages.

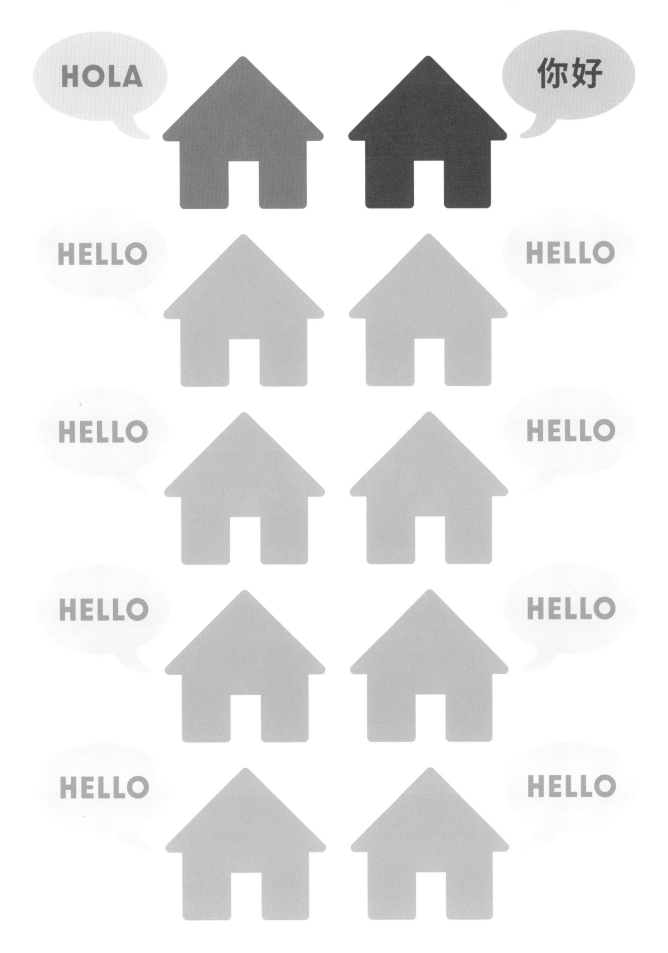

THE PLAIN OLD WEIRD

THE ODDS OF SOMEONE BEING BORN WITH 11 FINGERS OR TOES: 1 IN 500 TO 1,000

This is known as polydactyly, and it seems like an amazingly common number of people, but the studies bear it out. Among people of Asian and Caucasian descent, there is a greater likelihood of an extra thumb, while for those of African descent, an extra small finger or toe is more common. The phenomenon is partially hereditary, and there is a much greater chance of offspring having an extra digit if both parents have them, too.

MORE THAN 10 PEOPLE A YEAR IN THE U.S. ARE KILLED BY A VENDING MACHINE

This sounds like some schlocky 1980s horror film (*Attack of the Killer Soda Machines from Hell*), but amazingly, it's true. These deaths don't involve said machines coming alive and chasing people through abandoned shopping malls (as far as we know!) but, rather, are caused by human activity. Because of course they are. Usually, it's because someone didn't get the item they wanted and proceeded to rock the machine back and forth to try to dislodge it. They rock too far, and the vending machine falls on them, and . . . well, you get the idea. Most machines now come with a warning not to rock them, and many are fastened to walls to prevent these kinds of unfortunate mishaps. You may not get that bag of potato chips, but at least you'll live.

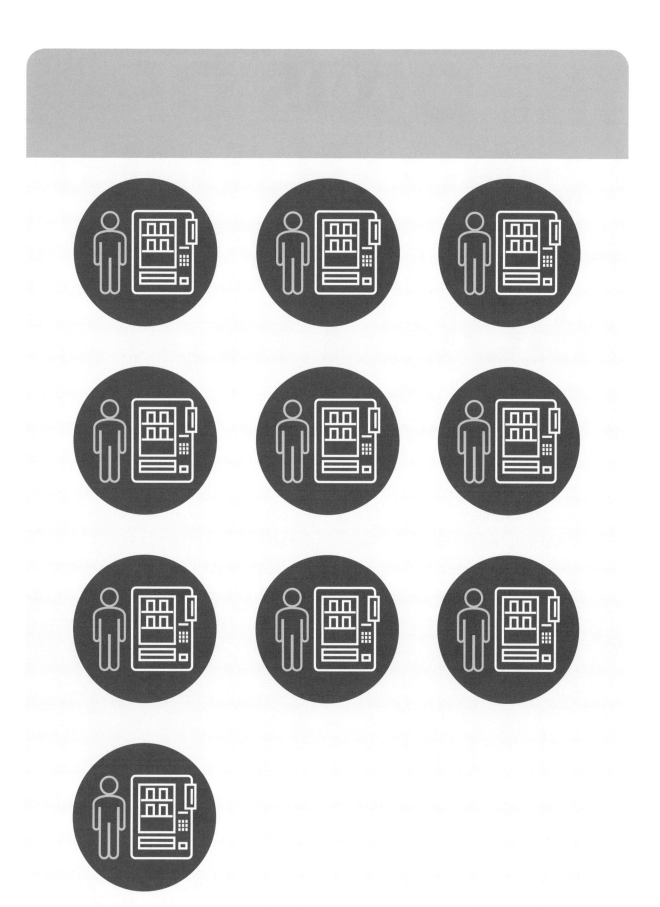

THE ODDS OF BEING INJURED BY A TOILET: ABOUT 1 IN 10,000

The CDC actually did a whole study about bathroom injuries, including the rather undignified toilet mishaps. It's actually more common than most people realize, and usually involves things like toilet seats breaking, forgetting to put down the seat before sitting, and so on. It's probably not surprising that bathroom injuries in general are fairly common (slipping in the bathtub or shower, for example, is easy to do), but toilet accidents are particularly embarrassing and happen to around 75,000 people in the U.S. every year.

1:10,000

GENERALLY, FEWER CHILDREN ARE BORN ON SUNDAY THAN ANY OTHER DAY

At least as far as studies from a few years ago have found. This isn't an overly scientific result that stays consistent from year to year, and there could be many different factors. Conversely, it seems like Tuesdays and Thursdays see the births of the most babies, though again, this is subject to change and no one really knows why.

MON	TUE	WED	THU	FRI	SAT	SUN

THE ODDS OF GETTING ATTACKED BY A SHARK WHEN SWIMMING IN THE OCEAN: ABOUT 1 IN 3.75 MILLION

Despite what movies like *Jaws* suggest, shark attacks against humans are very rare. In fact, the odds of dying in a car accident are a mere 1 in 63, and yet we drive without giving it a moment's thought. Despite what you may have heard, sharks are not likely to confuse most divers and swimmers with seals, and swimming in a group of people will greatly reduce the risk of being attacked, or even approached. So the next time you hit the beach you can relax and not worry that a great white is out there just waiting for a tasty snack (i.e., you).

1:3.75 MILLION

IN THE U.S., THERE ARE NEARLY 12,000 ANNUAL INJURIES RELATED TO FALLING TVS

This stat is as of 2017, and it's possible that the number, 11,800 to be precise, has only increased since then. The Consumer Product Safety Commission looked at a wide range of injuries, which involved TVs falling off of furniture (everything from tables to dressers). They didn't even consider the hazards of wall-mounted TVs (that will likely require another study). Further, the study only looked at nonfatal examples. A good number of these accidents consisted of thin TVs tipping over and falling, or even the whole piece of furniture tipping over. When TVs were placed on TV-appropriate furniture (i.e., pieces specifically designed to hold TVs), the number of accidents went way down. So the takeaway should be to not try to balance your TV on something not meant to hold it.

 = 1,000 injuries

1 IN 4 AMERICANS BUY VALENTINE'S DAY GIFTS FOR THEIR PETS

Yes, nearly 27% of people have responded to surveys saying that they will spend something on their beloved animal companion(s) for Valentine's Day, at an average of just over $12 each. Maybe bearded dragons love getting cards? In addition, nearly half of people polled have said that they have or will plan something special to do on the day, even if it's just taking their dog for a walk or spending extra time with the cat, or the iguana. Clearly, we're devoted to our pets, and for many, they're probably better companions than those dumb old humans, anyway.

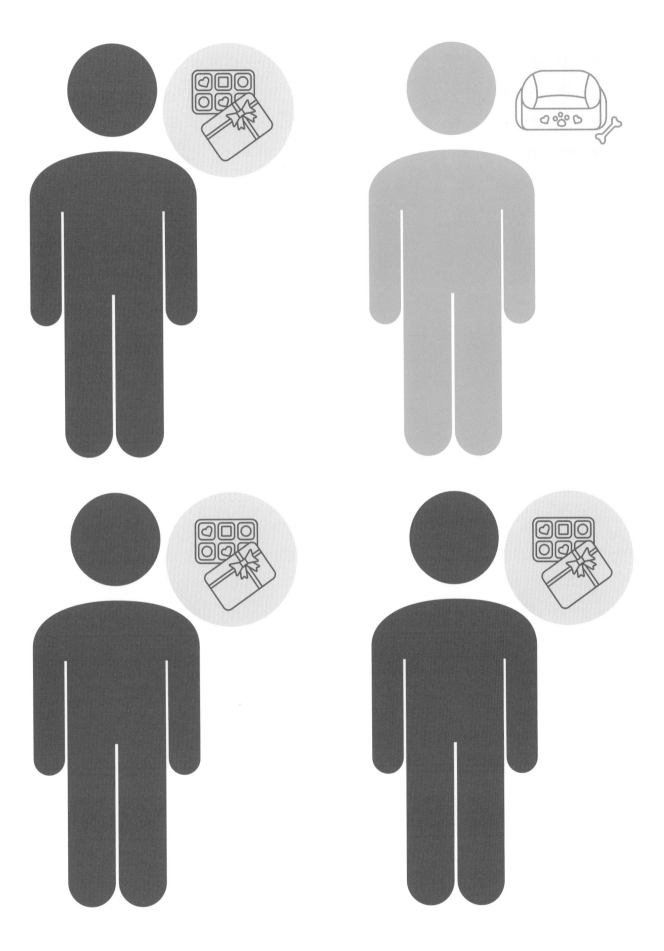

ABOUT 100 PEOPLE EVERY YEAR CHOKE TO DEATH ON PEN CAPS

These deaths are not just children. It seems most of us don't grow out of the oral stage (that is, if you buy Freud's ideas). People seem to love to chew on things, like pen caps, and this can be a dangerous problem. Chewing can lead to accidental swallowing, which can lead to accidental death. So, about 100 people still die every year from this unfortunate fate, but the number used to be higher. You may have noticed that a good number of pen caps now have a hole in the narrow end. Pen maker BIC realized that by making the hole prominent, it could allow air to get through to someone if the cap became lodged in their windpipe. It was a simple innovation that has definitely saved lives, and other manufacturers have followed suit.

THE ODDS THAT YOU WILL DIE IN AN AIRPLANE CRASH: ABOUT 1 IN 1 MILLION

This number should come as a relief to those with a fear of flying, but it also comes with some qualifications. There is no one average number because some people fly frequently, while others barely fly at all. Someone who flies eight times a year on business trips will have different odds than someone who steps onto a plane once every three years for a vacation. Frequent flyers have a greater risk simply because they are in the air more often. Again, there is still a much greater risk of dying in an automobile accident, but we tend to believe cars are safer because we use them every day and we are in control, instead of relinquishing control of the vehicle to pilots. But flying remains a very safe form of travel.

1:1 MILLION

PILLOWS, BEDS, AND MATTRESSES ACCOUNT FOR ABOUT 816,000 INJURIES EACH YEAR

While a large number or these involve infants and toddlers, as well as the elderly, a surprising number of accidents and injuries befall teenagers and adults, too. These can include everything from falling out of bed to a bed frame collapsing, to sleeping badly on a pillow and injuring the neck, and dozens of other causes you've never even thought about, but will now proceed to worry about every night as you try to fall asleep. It turns out that even your bed isn't necessarily a safe place!

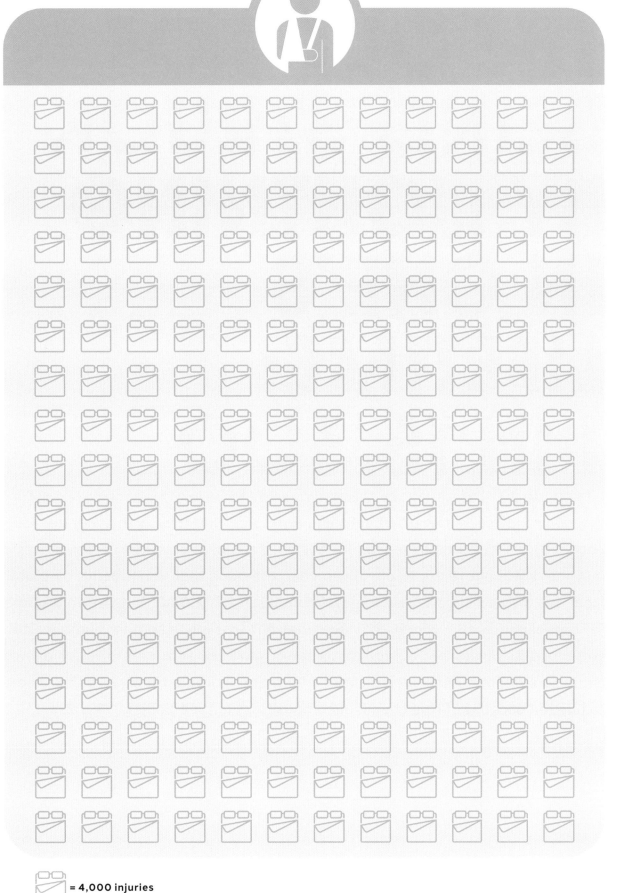

= 4,000 injuries

THE ODDS OF BEING STRUCK BY LIGHTNING IN ONE'S LIFETIME: 1 IN 15,300

We've already seen how often lightning strikes around the Earth every day (see page 90), so it stands to reason that there's a reasonable risk of being hit by it, right? Well, not really. In any given year, the odds that you'll be hit by a bolt from out of the blue are very remote, about 1 in 1,222,000. But over a lifetime (of approximately 80 years), that averages out to a much smaller number. Still, you probably don't need to walk around worrying about it, unless you're one of those rare people that's been struck by lightning multiple times, and this seems to happen because of rare chemical compositions in their bodies.

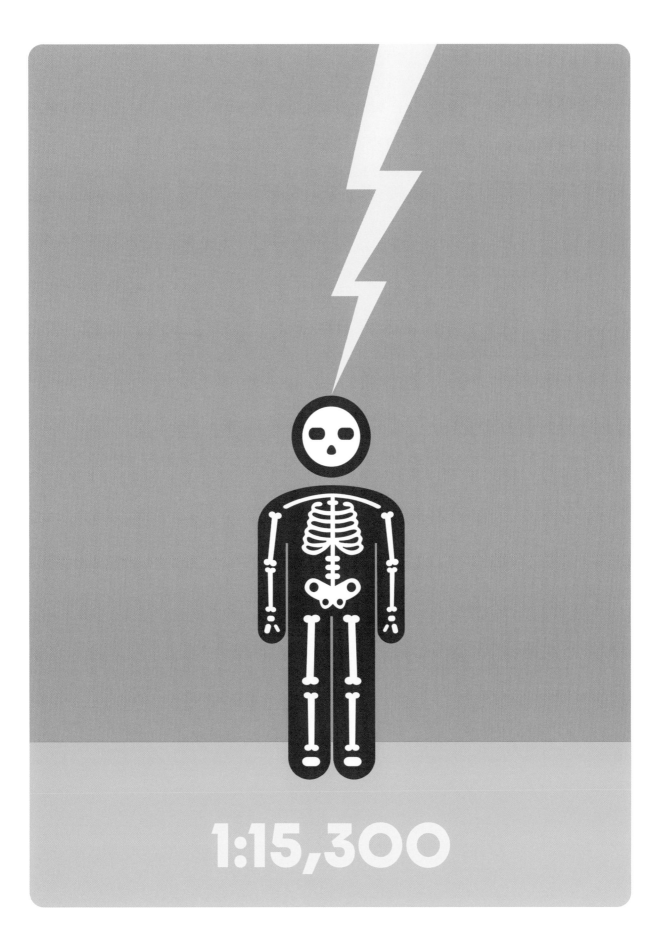

FAR MORE PEOPLE ARE KILLED BY COWS EVERY YEAR THAN SHARKS

Shark attacks are rare (see page 154) and probably only result in 1 to 5 deaths a year. By that measure, cows are way more deadly. It's estimated that cows kill about 20 people a year. They are generally docile and peaceful creatures, but they might charge if they feel threatened, or if they panic. Walkers who cross fields filled with cows, and especially bulls, can avoid most trouble by simply minding their own business and keeping their distance. Keep dogs on leashes around cows and don't upset them, and you're unlikely to join this exclusive and unfortunate club.

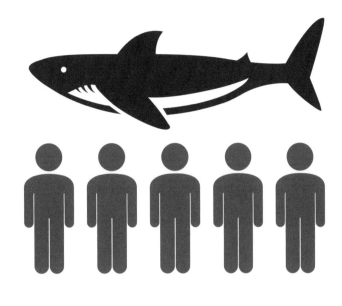

BONUS: 73.6% OF ALL STATISTICS ARE MADE UP

Realistically, this number is probably not true, so take it with a grain of salt. There's a greater than 50% chance that it's not true at all.

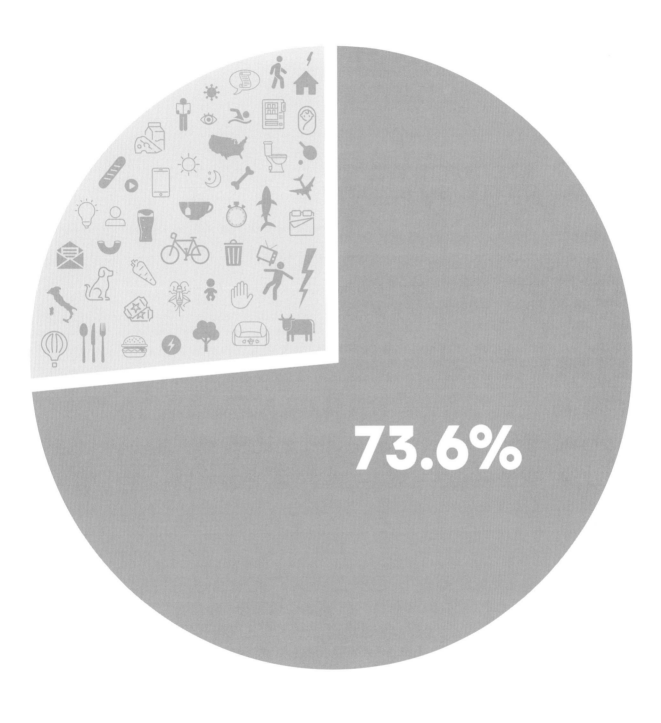

73.6%

SOURCES

THE ONLINE WORLD

1 Facebook and Backlink
2 Internet World Stats
3 Internet Live Stats, Orbit Media Studios
4 Statista
5 Statista
6 Statcounter
7 Statista
8 Internet Live Stats
9 Federal Communications Commission
10 Statista, DataProt, and Tech Jury
11 Statista
12 Drive Research
13 Pew Research Center
14 Pew Research Center
15 Intel

EAT, DRINK, LIVE

1 American Farm Bureau Federation
2 American Farm Bureau Federation
3 The Local (France)
4 The Culture Trip
5 Russian Agricultural Bank and the Center for Industry Expertise
6 American Farm Bureau Federation
7 The Drinks Business and Drinkaware
8 The U.S. Alcohol and Tobacco Tax and Trade Bureau, the Commerce Department, the U.S. Census, and the National Beer Wholesalers Association Industry Affairs
9 Petfood Industry
10 The Local (France)
11 Logitech
12 Themed Entertainment Association (TEA)
13 Sleepopolis
14 World Economic Forum
15 Statista
16 Statista
17 The Fiscal Times
18 Mental Floss and Statista
19 American Federation of Astrologers and *Scientific American*

SCIENCE

GLOBAL FACTS AND FIGURES

THE PLAIN OLD WEIRD

ABOUT THE AUTHOR

Tim Rayborn has written a large number of books and magazine articles (more than thirty each!), especially on subjects such as music, the arts, general knowledge, and history, though none of them are nearly as good as the classics in this book! He will no doubt write more. He lived in England for many years and studied at the University of Leeds, which means he likes to pretend that he knows what he's talking about. Incidentally, Statistics is the only college math class he got an A in, for what it's worth.

He's also an almost-famous musician who plays dozens of unusual instruments from all over the world that most people have never heard of and usually can't pronounce.

He has appeared on more than forty recordings, and his musical wanderings and tours have taken him across the U.S., all over Europe, to Canada and Australia, and to such romantic locations as Umbrian medieval towns, Marrakech, Vienna, Renaissance chateaux, medieval churches, and high school gymnasiums.

He currently lives in Northern California with many books, recordings, and instruments, and a sometimes-demanding cat. He's pretty enthusiastic about good wines and cooking excellent food.

timrayborn.com

ABOUT CIDER MILL PRESS BOOK PUBLISHERS

Good ideas ripen with time. From seed to harvest, Cider Mill Press brings fine reading, information, and entertainment together between the covers of its creatively crafted books. Our Cider Mill bears fruit twice a year, publishing a new crop of titles each spring and fall.

"Where Good Books Are Ready for Press"

Visit us online at
cidermillpress.com

or write to us at
PO Box 454
12 Spring St.
Kennebunkport, Maine 04046